WHY DO ANIMALS DO THAT?

WHY DO ANIMALS NEED A MATE?

BY ROSIE BANKS

Gareth Stevens
PUBLISHING

Please visit our website, www.garethstevens.com. For a free color catalog of all our high-quality books, call toll free 1-800-542-2595 or fax 1-877-542-2596.

Library of Congress Cataloging-in-Publication Data

Names: Banks, Rosie, 1978- author.
Title: Why do animals need a mate? / Rosie Banks.
Description: New York : Gareth Stevens Publishing, [2024] | Series: Why do animals do that? | Includes bibliographical references and index.
Identifiers: LCCN 2022051775 (print) | LCCN 2022051776 (ebook) | ISBN 9781538285244 (library binding) | ISBN 9781538285237 (paperback) | ISBN 9781538285251 (ebook)
Subjects: LCSH: Courtship in animals–Juvenile literature. | Sexual behavior in animals–Juvenile literature.
Classification: LCC QL761 .B34 2024 (print) | LCC QL761 (ebook) | DDC 591.56/2–dc23/eng/20221107
LC record available at https://lccn.loc.gov/2022051775
LC ebook record available at https://lccn.loc.gov/2022051776

Published in 2024 by
Gareth Stevens Publishing
2544 Clinton Street
Buffalo, NY 14224

Designer: Claire Wrazin
Editor: Therese Shea

Photo credits: Cover Signature Message/Shutterstock.com; Series Art (texture) Anna Timoshenko; p. 5 Deo/Shutterstock.com; p. 7 Ondrej Prosicky/Shutterstock.com; p. 9 colourbird/Shutterstock.com; p. 9 (inset) Bonnie Taylor Barry/Shutterstock.com; p. 11 Martin Mecnarowski/Shutterstock.com; p. 13 Julee75/Shutterstock.com; p. 15 Julian Wiskemann/Shutterstock.com; p. 17 Nagel Photography/Shutterstock.com; p. 17 (inset) Tetyana Dotsenko/Shutterstock.com; p. 19 Sergey Uryadnikov/Shutterstock.com; p. 20 Bidru/Shutterstock.com.

Printed in the United States of America

CPSIA compliance information: Batch #CSGS24: For further information contact Gareth Stevens, New York, New York at 1-800-542-2595.

CONTENTS

Boldface words appear in the glossary.

Better Together

Have you ever seen a pair of animals in the park or at the zoo? Maybe they play, **snuggle**, or chase each other! That pair might be mates. Animals that come together to reproduce, or have young, are called mates.

BALD EAGLE MATES

Meeting Their Match

Mates help each other **survive**. They help their **species** survive too. Most animals need another animal of their species to have babies, or offspring. If these animals didn't find mates, their species would die out! An animal looks for the best mate possible.

KING PENGUINS

Only the Best

Animals choose mates that have certain **qualities**. The **males** of many bird species have bright feathers to attract, or draw, **females**. For peacocks, males with the longest feathers are chosen as mates. A mate must also be healthy.

cardinal mates

Good qualities in a mate are different for different animal species. For some deer, females are more likely to choose males with large **antlers**. Some male bats give off a special odor, or smell, to attract a mate.

Family Matters

After animals mate, young may grow in the mother's body. Mother animals may give birth to live young. These babies need care. Sometimes, both mates care for the young. For example, **meerkat** fathers help feed and guard their offspring.

MEERKATS

Some animals lay eggs. Both parents or just one may care for the eggs. Sometimes, a mother lays the eggs and leaves. The young must survive without help. For example, green sea turtles don't take care of their eggs or young.

BABY SEA TURTLES

Together Forever?

Some animals have many mates during their lives. Others have one. Gray wolves, bald eagles, and macaroni penguins are species that mate for life. These mates share food and work, such as caring for their young, their whole lives.

GRAY WOLVES

macaroni penguins

Single Parent

Most animals need a mate to have young. A small number don't, such as some bees, ants, snakes, fish, and lizards. These animals sometimes produce offspring alone. Komodo dragons are an example. Females may lay eggs even if no mates are around.

KOMODO DRAGON

Keep Learning!

So many animal species choose mates in interesting ways. Gibbons are small apes. Each has a call to attract a mate. Mates join their calls into a song. They sing it to each other their whole lives! How does your favorite animal find its mate?

gibbons

ANIMALS MAY NEED MATES:

- To reproduce
- To care for eggs or young
- To guard family
- To build and guard homes
- To find and share food

GLOSSARY

antler: A bony, branched growth on the head of an animal.

female: An animal that is able to carry young or produce eggs.

male: An animal that produces cells that join together with eggs in order to reproduce.

meerkat: A small mongoose found in the deserts and grasslands of southern Africa.

quality: Feature.

snuggle: To sit or lie close to another for comfort.

species: A group of plants or animals that are all of the same kind.

survive: To live through something.

FOR MORE INFORMATION

BOOKS

Jenkins, Steve, and Robin Page. *Look at Me! How to Attract Attention in the Animal World*. Boston, MA: Houghton Mifflin Harcourt, 2018.

Lindeen, Mary. *Animal Mothers and Fathers.* Chicago, IL: Norwood House Press, 2018.

WEBSITES

6 Animal Species That Mate for Life
www.britannica.com/list/6-animal-species-that-mate-for-life
Read about animal species that choose mates for their whole life.

To Breed or Not to Breed
askabiologist.asu.edu/mate-choice#
Find out how different animals attract their mates.

INDEX